THE EPI DIET BIBLE

I0514552

Beginners Guide To A Well-Balanced Diet For Exocrine Pancreatic Insufficiency: Recipes And Nutritional Advice

CRUE GAGE

Table of Contents

Introductory

Exocrine Pancreatic Insufficiency (EPI) is a condition in which the pancreas fails to generate an adequate quantity of digestive enzymes to facilitate the digestion of food. In order to facilitate the absorption of lipids, proteins, and carbohydrates into the bloodstream, the pancreas typically secretes enzymes such as lipase, protease, and amylase into the small intestine.
The absence of these enzymes results in the malabsorption of nutrients in individuals with EPI.

Diarrhea, weight loss, bloating, abdominal discomfort, and deficiencies in fat-soluble vitamins such as A, D, E, and K are all potential symptoms of this condition. EPI can be the consequence of a variety of conditions that affect the pancreas, including cystic fibrosis, chronic pancreatitis, pancreatic cancer, or surgical excision of a portion of the pancreas.

Enzyme replacement therapy is the standard treatment, which entails the administration of synthetic pancreatic enzymes with meals to facilitate digestion and enhance nutrient assimilation. The quality of life for individuals with EPI can be enhanced through proper management under medical supervision, which can alleviate symptoms.

Diet plays a crucial role in managing Exocrine Pancreatic Insufficiency (EPI). Here are some key aspects:

- **Nutrient Absorption**: With EPI, the body struggles to absorb fats, proteins, and carbohydrates. A well-planned diet can help ensure adequate intake of essential nutrients.

- **Low-Fat Diet**: Reducing dietary fat can help minimize symptoms like diarrhea and abdominal discomfort, especially if enzyme replacement therapy is not fully effective.

- **Frequent, Small Meals**: Eating smaller, more frequent meals can aid digestion and reduce stress on the digestive system.

- **High-Quality Protein Sources**: Incorporating easily digestible protein sources, such as lean meats, fish, eggs, and dairy, can help meet protein needs.

- **Supplementation**: Supplementing with fat-soluble vitamins (A, D, E, K) may be necessary, as malabsorption can lead to deficiencies.

- **Avoiding Certain Foods**: Some people may find relief by avoiding high-fiber foods, spicy dishes, and highly processed foods that can exacerbate digestive symptoms.

- **Hydration**: Staying hydrated is important, especially if diarrhea is present, to prevent dehydration.

Working with a healthcare provider or dietitian can help tailor a diet that addresses individual needs and symptoms effectively.

CHAPTER ONE

Causes And Symptoms

Causes of Exocrine Pancreatic Insufficiency (EPI):

- **Chronic Pancreatitis**: Inflammation of the pancreas can damage the cells responsible for enzyme production.

- **Cystic Fibrosis**: This genetic disorder affects the pancreas, leading to thick secretions that obstruct enzyme flow.

- **Pancreatic Cancer**: Tumors can interfere with enzyme production or obstruct the pancreatic duct.

- **Pancreatic Surgery**: Surgical removal of part or all of the pancreas can reduce enzyme output.

- **Diabetes**: Sometimes, diabetes, particularly type 1, can be associated with EPI due to autoimmune damage to the pancreas.

- **Other Conditions**: Conditions like Crohn's disease, certain infections, or celiac disease can also contribute to EPI.

Symptoms of EPI:

- **Diarrhea**: Often oily or foul-smelling, due to unabsorbed fats.

- **Weight Loss**: Unintentional weight loss due to malabsorption of nutrients.

- **Abdominal Pain**: Discomfort or cramping, which may be related to digestion issues.

- **Bloating and Gas**: Increased gas production and bloating can occur.

- **Fatty Stools (Steatorrhea)**: Stools may be pale, greasy, and difficult to flush due to high fat content.

- **Nutrient Deficiencies**: Signs of deficiencies in vitamins and minerals, particularly fat-soluble vitamins (A, D, E, K).

If someone experiences these symptoms, it's important to consult a healthcare provider for proper diagnosis and management.

Macronutrients: Proteins, Fats, And Carbohydrates

Macronutrients are the nutrients that provide energy and are essential for growth and development. They include proteins, fats, and carbohydrates. Here's a brief overview of each:

1. Proteins:

- **Function**: Essential for building and repairing tissues, making enzymes and hormones, and supporting immune function.

- **Sources**: Meat, fish, eggs, dairy products, legumes, nuts, and seeds.

- **Recommended Intake**: Generally, 10-35% of total daily calories, but needs can vary based on activity level and health status.

2. Fats:

- **Function**: Provide a concentrated source of energy, aid in the absorption of fat-soluble vitamins (A, D, E, K), and are important for cell structure and hormone production.

- **Sources**: Oils, butter, avocados, nuts, seeds, and fatty fish. It's important to focus on healthy fats, such as unsaturated fats, while limiting saturated and trans fats.

- **Recommended Intake**: Typically, 20-35% of total daily calories.

3. Carbohydrates:

- **Function**: The body's primary source of energy, especially for the brain and muscles during high-intensity exercise.

They also provide fiber, which is important for digestive health.

- **Sources**: Fruits, vegetables, grains, legumes, and dairy products. Complex carbohydrates (like whole grains) are generally preferred over simple sugars.

- **Recommended Intake**: Usually, 45-65% of total daily calories.

In managing EPI, balancing these macronutrients while considering enzyme replacement therapy can help optimize digestion and nutrient absorption.

Micronutrients: Vitamins And Minerals

Micronutrients are essential nutrients required by the body in smaller amounts compared to macronutrients. They include vitamins and minerals, both of which play crucial roles in various bodily functions.

Vitamins:

Fat-Soluble Vitamins:

- **Vitamin A**: Important for vision, immune function, and skin health. Sources include carrots, sweet potatoes, and leafy greens.
- **Vitamin D**: Crucial for bone health and immune function. Sources include sunlight, fatty fish, and fortified dairy products.

- **Vitamin E**: Acts as an antioxidant and supports immune function. Sources include nuts, seeds, and vegetable oils.
- **Vitamin K**: Essential for blood clotting and bone health. Sources include leafy greens and broccoli.

Water-Soluble Vitamins:

- **Vitamin C**: Important for collagen synthesis, antioxidant protection, and immune function. Sources include citrus fruits, berries, and bell peppers.
- **B Vitamins** (e.g., B1, B2, B3, B6, B12, folate): Involved in energy metabolism, brain function, and red blood cell production. Sources include whole grains, meat, eggs, and legumes.

Minerals:

Macro-Minerals:

- **Calcium**: Essential for bone health and muscle function. Sources include dairy products, leafy greens, and fortified foods.

- **Potassium**: Important for fluid balance, muscle contractions, and nerve function. Sources include bananas, potatoes, and beans.

- **Magnesium**: Supports muscle and nerve function and is involved in over 300 biochemical reactions. Sources include nuts, seeds, whole grains, and leafy greens.

Trace Minerals:

- **Iron**: Essential for oxygen transport in the blood. Sources include red meat, beans, and fortified cereals.
- **Zinc**: Important for immune function and wound healing. Sources include meat, shellfish, legumes, and seeds.
- **Selenium**: Acts as an antioxidant and supports thyroid function. Sources include Brazil nuts, seafood, and eggs.

Micronutrients are vital for maintaining overall health, and deficiencies can lead to various health issues. For individuals with EPI, special attention should be given to fat-soluble vitamins and minerals to ensure adequate absorption and avoid deficiencies.

CHAPTER TWO

The Role Of Enzyme Replacement Therapy

Enzyme Replacement Therapy (ERT) is a critical treatment for individuals with Exocrine Pancreatic Insufficiency (EPI). Here's how it works and its importance:

Role of Enzyme Replacement Therapy:

- **Restores Digestive Function**: ERT provides the digestive enzymes that the pancreas is unable to produce in sufficient quantities. These enzymes help break down fats, proteins, and carbohydrates, facilitating proper digestion and nutrient absorption.

- **Improves Symptoms**: By aiding digestion, ERT can significantly reduce symptoms associated with EPI, such as diarrhea, bloating, abdominal pain, and fatty stools (steatorrhea).

- **Prevents Nutritional Deficiencies**: With effective enzyme replacement, the body can better absorb essential nutrients, reducing the risk of deficiencies in vitamins and minerals, particularly fat-soluble vitamins (A, D, E, K).

- **Customizable Dosage**: The dosage of enzyme supplements can be adjusted based on individual needs, dietary fat intake, and symptoms, ensuring optimal management of the condition.

- **Enhances Quality of Life**: By managing symptoms and improving nutrient absorption, ERT can significantly enhance overall health and quality of life for individuals with EPI.

Administration:

- **Timing**: Enzymes are typically taken with meals and snacks to coincide with the digestive process.

- **Form**: Enzyme supplements are available in various forms, including capsules and tablets, and they may contain varying amounts of enzymes.

Regular follow-ups with a healthcare provider are essential to monitor the effectiveness of therapy, adjust dosages, and address any potential side effects or complications.

Essential Foods To Include & Avoid

Here are some essential foods to include and avoid for managing Exocrine Pancreatic Insufficiency (EPI):

Essential Foods to Include:

Lean Proteins:

• Skinless poultry, fish, eggs, and low-fat dairy products are easier to digest and provide necessary protein.

Healthy Fats:

• Avocados, olive oil, and nuts in moderation can provide healthy fats while being mindful of portion sizes, as fat digestion may still be challenging.

Complex Carbohydrates:

• Whole grains (brown rice, quinoa, oats), fruits, and vegetables provide energy and fiber, which can support digestive health.

Low-Fiber Foods:

• Some people with EPI may tolerate lower-fiber options better, such as white rice or cooked vegetables, to reduce bloating.

Fat-Soluble Vitamins:

• Foods rich in vitamins A, D, E, and K (like leafy greens, nuts, and fish) can help address potential deficiencies.

Foods to Avoid

High-Fat Foods:

- Fried foods, fatty cuts of meat, and full-fat dairy products can be harder to digest and may exacerbate symptoms.

Processed Foods:

- Many processed foods are high in unhealthy fats and sugars, which can lead to digestive discomfort.

High-Fiber Foods:

- While fiber is important, high-fiber foods (like whole nuts, raw vegetables, and certain whole grains) may worsen bloating and gas in some individuals.

Spicy Foods:

- Spices can irritate the digestive tract and may lead to discomfort.

Caffeine and Alcohol:

- Both can irritate the digestive system and may exacerbate symptoms.

General Tips:

- **Small, Frequent Meals**: Eating smaller meals throughout the day can help manage symptoms.

- **Stay Hydrated**: Drinking plenty of water is essential, especially if experiencing diarrhea.

Consulting with a healthcare provider or dietitian can help tailor dietary choices to individual needs and preferences.

Portion Control And Meal Timing

Portion control and meal timing are important aspects of managing Exocrine Pancreatic Insufficiency (EPI) effectively. Here's how they can help:

Portion Control:

- **Smaller Portions**: Eating smaller, more frequent meals can ease the digestive process, reducing the workload on the pancreas and minimizing symptoms such as bloating and discomfort.

- **Mindful Eating**: Paying attention to hunger and fullness cues can prevent overeating and help maintain a balanced diet.

- **Balanced Plates**: Aim for a balanced intake of proteins, fats, and carbohydrates in each meal, adjusting portions based on individual tolerance and nutritional needs.

Meal Timing:

- **Regular Meal Schedule**: Establishing a consistent meal schedule can help regulate digestion and enzyme production. This can include three main meals and two to three snacks throughout the day.

- **Pre-Meal Enzymes**: Taking enzyme supplements just before meals ensures they are available to aid in digestion when food is consumed.

- **Avoid Late Meals**: Eating earlier in the evening can help reduce discomfort during digestion and promote better sleep.

- **Timing with Activity**: Plan meals and snacks around physical activity to ensure adequate energy levels and avoid discomfort during exercise.

By focusing on portion control and meal timing, individuals with EPI can optimize digestion, enhance nutrient absorption, and improve overall quality of life. Working with a healthcare provider or dietitian can further personalize these strategies.

CHAPTER THREE
7 Days Sample Meal Plans

Here's a 7-day sample meal plan for managing Exocrine Pancreatic Insufficiency (EPI). Each day includes three meals and two snacks, focusing on easily digestible foods and balanced macronutrients. Adjust portion sizes as needed based on individual tolerance.

Day 1:

- **Breakfast**: Scrambled eggs with spinach and a slice of whole-grain toast
- **Snack**: Greek yogurt (low-fat) with honey
- **Lunch**: Grilled chicken breast with quinoa and steamed broccoli
- **Snack**: A small handful of almonds

- **Dinner**: Baked salmon with sweet potato and sautéed green beans

Day 2:

- **Breakfast**: Oatmeal made with low-fat milk, topped with banana slices
- **Snack**: Cottage cheese with pineapple
- **Lunch**: Turkey and avocado wrap (using a whole-grain tortilla) with a side of carrot sticks
- **Snack**: Rice cakes with a thin layer of peanut butter
- **Dinner**: Stir-fried tofu with bell peppers and brown rice

Day 3:

- **Breakfast**: Smoothie with spinach, banana, and low-fat yogurt

- **Snack**: A small apple with cheese
- **Lunch**: Lentil soup with a side salad (dressing on the side)
- **Snack**: Sliced cucumber with hummus
- **Dinner**: Grilled shrimp tacos with corn tortillas and a side of black beans

Day 4:

- **Breakfast**: Scrambled eggs with diced tomatoes and a slice of whole-grain toast
- **Snack**: A small bowl of mixed berries
- **Lunch**: Quinoa salad with chickpeas, cucumbers, and feta cheese
- **Snack**: A small handful of walnuts

- **Dinner**: Baked chicken thighs with roasted vegetables (carrots and zucchini)

Day 5:

- **Breakfast**: Yogurt parfait with granola (low sugar) and berries
- **Snack**: Celery sticks with light cream cheese
- **Lunch**: Grilled fish tacos with cabbage slaw (on corn tortillas)
- **Snack**: Baby carrots with hummus
- **Dinner**: Turkey meatballs with whole wheat pasta and marinara sauce

Day 6:

- **Breakfast**: Smoothie with mixed fruits and almond milk
- **Snack**: Hard-boiled egg

- **Lunch**: Spinach salad with grilled chicken, strawberries, and vinaigrette
- **Snack**: Rice cakes with avocado
- **Dinner**: Baked tilapia with quinoa and steamed asparagus

Day 7:

- **Breakfast**: Whole-grain pancakes with a drizzle of maple syrup and sliced bananas
- **Snack**: A small pear
- **Lunch**: Vegetable soup with a side of whole-grain bread
- **Snack**: Low-fat yogurt with a sprinkle of cinnamon
- **Dinner**: Stuffed bell peppers with ground turkey and brown rice

Tips:

- Adjust portion sizes based on individual needs and tolerance.
- Stay hydrated throughout the day.
- Consider taking enzyme supplements with meals as advised by a healthcare provider.

Always consult with a healthcare provider or dietitian to tailor meal plans to specific dietary needs and preferences.

Breakfast Recipes

Here are some simple and delicious breakfast recipes that are suitable for managing Exocrine Pancreatic Insufficiency (EPI):

1. Scrambled Eggs with Spinach:

Ingredients:

- 2 eggs
- 1 cup fresh spinach
- Salt and pepper to taste
- 1 teaspoon olive oil or butter

Instructions:

- Heat olive oil or butter in a non-stick skillet over medium heat.
- Add spinach and sauté until wilted.

- Beat eggs in a bowl, season with salt and pepper, then pour into the skillet.
- Stir gently until eggs are cooked to your desired consistency. Serve warm.

2. Oatmeal with Banana:

Ingredients:

- 1/2 cup rolled oats
- 1 cup low-fat milk (or almond milk)
- 1 banana, sliced
- 1 teaspoon honey (optional)
- A sprinkle of cinnamon (optional)

Instructions:

- In a pot, bring milk to a boil.

- Stir in oats and reduce heat. Cook for about 5 minutes, stirring occasionally.
- Remove from heat, top with banana slices, honey, and cinnamon. Serve warm.

3. Smoothie with Spinach and Yogurt:

Ingredients:

- 1 cup fresh spinach
- 1 banana
- 1/2 cup low-fat yogurt
- 1/2 cup almond milk (or any milk)
- 1 tablespoon honey (optional)

Instructions:

- Combine spinach, banana, yogurt, and milk in a blender.

- Blend until smooth. Adjust consistency with more milk if needed.
- Sweeten with honey if desired. Serve immediately.

4. Yogurt Parfait:

Ingredients:

- 1 cup low-fat yogurt
- 1/2 cup granola (low sugar)
- 1/2 cup mixed berries (fresh or frozen)

Instructions:

- In a glass or bowl, layer yogurt, granola, and berries.
- Repeat layers if desired. Enjoy immediately.

5. Whole-Grain Pancakes:

Ingredients:

- 1 cup whole-grain flour
- 1 tablespoon baking powder
- 1 tablespoon sugar (optional)
- 1 cup low-fat milk
- 1 egg
- 1 tablespoon melted butter or oil

Instructions:

- In a bowl, mix flour, baking powder, and sugar.
- In another bowl, whisk together milk, egg, and melted butter.
- Pour wet ingredients into dry ingredients and stir until just combined.
- Heat a non-stick skillet over medium heat and pour batter to

form pancakes. Cook until bubbles form, then flip and cook until golden brown.

- Serve warm with a drizzle of maple syrup or fresh fruit.

These recipes focus on easily digestible ingredients while providing balanced nutrition. Adjust ingredients as needed based on individual preferences and dietary guidelines.

Lunch Recipes

Here are some lunch recipes suitable for managing Exocrine Pancreatic Insufficiency (EPI):

1. Grilled Chicken Quinoa Salad

Ingredients:

- 1 cup cooked quinoa

- 1 grilled chicken breast, sliced
- 1 cup cherry tomatoes, halved
- 1/2 cucumber, diced
- 1/4 cup feta cheese (optional)
- 2 tablespoons olive oil
- 1 tablespoon lemon juice
- Salt and pepper to taste

Instructions:

- In a large bowl, combine quinoa, grilled chicken, cherry tomatoes, cucumber, and feta.
- In a small bowl, whisk together olive oil, lemon juice, salt, and pepper.
- Drizzle dressing over the salad and toss gently. Serve chilled or at room temperature.

2. Turkey and Avocado Wrap:

Ingredients:

- 1 whole-grain tortilla
- 4-6 slices of turkey breast
- 1/4 avocado, sliced
- Handful of spinach or lettuce
- 1 tablespoon mustard or hummus (optional)

Instructions:

- Spread mustard or hummus on the tortilla if using.
- Layer turkey slices, avocado, and spinach on the tortilla.
- Roll tightly and slice in half. Serve with carrot sticks or a side salad.

3. Lentil Soup:

Ingredients:

- 1 cup lentils (rinsed)
- 1 onion, diced
- 2 carrots, diced
- 2 celery stalks, diced
- 4 cups vegetable or chicken broth
- 1 teaspoon cumin
- Salt and pepper to taste
- Olive oil

Instructions:

- In a pot, heat olive oil over medium heat. Add onion, carrots, and celery, sautéing until soft.
- Stir in lentils, broth, cumin, salt, and pepper. Bring to a boil.

- Reduce heat and simmer for 30-40 minutes until lentils are tender. Serve warm.

4. Spinach and Feta Stuffed Peppers:

Ingredients:

- 2 bell peppers, halved and seeds removed
- 1 cup cooked brown rice or quinoa
- 1 cup fresh spinach, chopped
- 1/2 cup feta cheese (optional)
- Salt and pepper to taste
- Olive oil

Instructions:

- Preheat oven to 375°F (190°C).
- In a bowl, mix cooked rice/quinoa, spinach, feta, salt, and pepper.

- Stuff the mixture into the halved bell peppers and place them in a baking dish.
- Drizzle with olive oil and cover with foil. Bake for 25-30 minutes, then uncover and bake for an additional 10 minutes. Serve warm.

5. Vegetable Stir-Fry with Tofu:

Ingredients:

- 1 block firm tofu, cubed
- 2 cups mixed vegetables (broccoli, bell peppers, carrots)
- 2 tablespoons soy sauce (low sodium)
- 1 tablespoon olive oil
- Cooked brown rice for serving

Instructions:

- In a pan, heat olive oil over medium heat. Add cubed tofu and cook until golden brown on all sides. Remove and set aside.
- In the same pan, add mixed vegetables and stir-fry until tender-crisp.
- Return tofu to the pan, add soy sauce, and stir well. Cook for another 2-3 minutes. Serve over brown rice.

These recipes are designed to be nutritious and easy to digest while supporting overall health. Adjust ingredients and portion sizes based on individual preferences and dietary needs.

Dinner Recipes

Here are some dinner recipes suitable for managing Exocrine Pancreatic Insufficiency (EPI):

1. Baked Salmon with Asparagus:

Ingredients:

- 2 salmon fillets
- 1 bunch asparagus, trimmed
- 2 tablespoons olive oil
- 1 lemon (sliced)
- Salt and pepper to taste

Instructions:

- Preheat the oven to 400°F (200°C).
- Place salmon fillets and asparagus on a baking sheet. Drizzle with olive oil and season with salt and pepper.

- Top salmon with lemon slices.
- Bake for 15-20 minutes, until salmon is cooked through and flakes easily with a fork. Serve warm.

2. Turkey Meatballs with Marinara Sauce:

Ingredients:

- 1 lb ground turkey
- 1/4 cup breadcrumbs (whole-grain)
- 1/4 cup grated Parmesan cheese
- 1 egg
- 1 teaspoon Italian seasoning
- 2 cups marinara sauce
- Cooked whole wheat pasta (optional)

Instructions:

- Preheat the oven to 375°F (190°C).
- In a bowl, combine ground turkey, breadcrumbs, Parmesan, egg, and Italian seasoning. Mix well and form into meatballs.
- Place meatballs on a baking sheet and bake for 20-25 minutes until cooked through.
- Meanwhile, heat marinara sauce in a saucepan. Once meatballs are done, add them to the sauce and simmer for a few minutes. Serve over pasta if desired.

3. Vegetable Stir-Fry with Chicken:

Ingredients:

- 1 lb chicken breast, sliced

- 2 cups mixed vegetables (bell peppers, broccoli, snap peas)
- 2 tablespoons soy sauce (low sodium)
- 1 tablespoon sesame oil (or olive oil)
- Cooked brown rice for serving

Instructions:

- Heat sesame oil in a large pan or wok over medium-high heat.
- Add sliced chicken and cook until browned and cooked through. Remove from pan.
- Add mixed vegetables to the pan and stir-fry for about 5-7 minutes until tender-crisp.
- Return chicken to the pan, add soy sauce, and stir well. Cook for an

additional 2 minutes. Serve over brown rice.

4. Stuffed Bell Peppers with Ground Beef and Rice:

Ingredients:

- 2 bell peppers, halved and seeds removed
- 1 lb lean ground beef
- 1 cup cooked brown rice
- 1 can diced tomatoes (14 oz)
- 1 teaspoon Italian seasoning
- Salt and pepper to taste

Instructions:

- Preheat the oven to 375°F (190°C).
- In a skillet, brown ground beef over medium heat. Drain excess fat.

- Add cooked rice, diced tomatoes, Italian seasoning, salt, and pepper. Mix well.
- Stuff the mixture into halved bell peppers and place them in a baking dish.
- Cover with foil and bake for 30-35 minutes. Serve warm.

5. Quinoa and Black Bean Bowl:

Ingredients:

- 1 cup cooked quinoa
- 1 can black beans (15 oz), rinsed and drained
- 1 cup corn (fresh or frozen)
- 1 avocado, diced
- 1 lime (juiced)
- Salt and pepper to taste
- Fresh cilantro (optional)

Instructions:

- In a large bowl, combine cooked quinoa, black beans, corn, and avocado.
- Drizzle with lime juice and season with salt and pepper. Toss gently to combine.
- Garnish with fresh cilantro if desired. Serve immediately.

These recipes are designed to be nutritious, easy to prepare, and gentle on the digestive system. Adjust ingredients and portion sizes based on individual preferences and dietary needs.

Snacks And Desserts Recipes

Here are some easy and nutritious snack and dessert recipes suitable for managing Exocrine Pancreatic Insufficiency (EPI):

Snacks:

1. Greek Yogurt with Honey and Berries:

Ingredients:

- 1 cup low-fat Greek yogurt
- 1 tablespoon honey
- 1/2 cup mixed berries (fresh or frozen)

Instructions:

- In a bowl, combine Greek yogurt and honey.
- Top with mixed berries. Serve chilled.

2. Rice Cakes with Peanut Butter and Banana:

Ingredients:

- 2 rice cakes
- 2 tablespoons peanut butter (or almond butter)
- 1 banana, sliced

Instructions:

- Spread peanut butter on each rice cake.
- Top with banana slices. Enjoy as a quick snack.

3. Hummus with Veggie Sticks:

Ingredients:

- 1 cup hummus
- Assorted vegetable sticks (carrots, cucumber, bell peppers)

Instructions:

- Serve hummus in a bowl with veggie sticks on the side for dipping. Enjoy!

Desserts:

1. Banana Oatmeal Cookies:

Ingredients:

- 2 ripe bananas, mashed
- 1 cup rolled oats
- 1/4 cup dark chocolate chips (optional)

Instructions:

- Preheat the oven to 350°F (175°C).
- In a bowl, mix mashed bananas and oats until well combined. Stir in chocolate chips if using.

- Drop spoonfuls of the mixture onto a baking sheet lined with parchment paper.
- Bake for 10-12 minutes or until golden. Let cool before serving.

2. Chia Seed Pudding:

Ingredients:

- 1/4 cup chia seeds
- 1 cup almond milk (or any milk)
- 1 tablespoon maple syrup (optional)
- 1/2 teaspoon vanilla extract

Instructions:

- In a bowl, combine chia seeds, almond milk, maple syrup, and vanilla extract. Stir well.
- Refrigerate for at least 4 hours or overnight until it thickens.

- Serve chilled, topped with fresh fruit or nuts if desired.

3. Baked Apples with Cinnamon:

Ingredients:

- 4 apples, cored
- 1/4 cup oats
- 1 teaspoon cinnamon
- 2 tablespoons honey or maple syrup

Instructions:

- Preheat the oven to 350°F (175°C).
- In a bowl, mix oats, cinnamon, and honey or maple syrup.
- Stuff the mixture into the cored apples.
- Place apples in a baking dish with a little water and bake for 20-25 minutes until tender. Serve warm.

These snacks and desserts are designed to be satisfying and gentle on the digestive system while providing essential nutrients. Adjust ingredients and portion sizes based on individual preferences and dietary needs.

CHAPTER FOUR

Managing Symptoms Through Diet

Managing symptoms of Exocrine Pancreatic Insufficiency (EPI) through diet involves making informed food choices that support digestion and nutrient absorption. Here are key strategies:

1. Balance Macronutrients:

• **Proteins**: Include lean sources like chicken, turkey, fish, and eggs, which are easier to digest.

• **Fats**: Focus on healthy fats (avocado, olive oil, nuts) but monitor portions, as high-fat meals can be harder to digest.

• **Carbohydrates**: Opt for complex carbohydrates (whole grains, fruits, vegetables) while being mindful of fiber intake to prevent bloating.

2. Choose Easily Digestible Foods

- **Cooked Vegetables**: Cooking makes vegetables easier to digest than raw options.

- **Low-Fiber Options**: If experiencing bloating, consider lower-fiber foods temporarily, like white rice or cooked carrots.

3. Portion Control:

- Eating smaller, more frequent meals can help reduce digestive strain and improve nutrient absorption.

4. Stay Hydrated

- Drink plenty of water throughout the day to support digestion and overall health.

5. Timing with Enzyme Replacement Therapy

- Take enzyme supplements with meals to aid in the digestion of fats, proteins, and carbohydrates. Adjust dosage based on meal size and fat content.

6. Limit Trigger Foods

- Identify and limit foods that exacerbate symptoms, such as high-fat, fried, or spicy foods, as well as caffeine and alcohol.

7. Incorporate Probiotics

- Foods like yogurt and kefir can support gut health, though individual tolerance may vary.

8. Monitor for Nutrient Deficiencies

- Regularly assess for deficiencies in fat-soluble vitamins (A, D, E, K) and minerals, and adjust diet or supplementation as needed.

9. Keep a Food Diary

- Tracking what you eat and your symptoms can help identify triggers and improve dietary choices.

By following these dietary strategies, individuals with EPI can manage symptoms effectively and enhance their overall quality of life. It's essential to work with a healthcare provider or dietitian to tailor these recommendations to individual needs.

Supplementation And Vitamins

For individuals with Exocrine Pancreatic Insufficiency (EPI), supplementation can play a crucial role in addressing nutritional deficiencies due to impaired digestion and absorption. Here's an overview of important supplements and vitamins:

1. Pancreatic Enzyme Replacement Therapy (PERT):

- **Description**: Enzymes (lipase, amylase, protease) that aid in the digestion of fats, carbohydrates, and proteins.

• **Use**: Taken with meals to improve nutrient absorption and alleviate symptoms like bloating and diarrhea.

2. Fat-Soluble Vitamins:

• Individuals with EPI are at risk of deficiencies in fat-soluble vitamins due to malabsorption:

Vitamin A:

- **Importance**: Supports vision, immune function, and skin health.
- **Sources**: Carrots, sweet potatoes, and dark leafy greens.

Vitamin D:

- **Importance**: Essential for bone health and immune function.
- **Sources**: Fatty fish, fortified dairy products, and sunlight exposure.

Vitamin E:

- **Importance**: Acts as an antioxidant and supports immune health.
- **Sources**: Nuts, seeds, and vegetable oils.

Vitamin K:

- **Importance**: Crucial for blood clotting and bone health.
- **Sources**: Leafy greens, broccoli, and Brussels sprouts.

3. B Vitamins:

- **Importance**: Involved in energy metabolism, red blood cell production, and brain health.
- **Sources**: Whole grains, meat, eggs, and legumes.

- **Note**: Supplementation may be necessary if dietary intake is insufficient.

4. Minerals:

- **Iron**: Important for oxygen transport in the blood. Sources include red meat, beans, and fortified cereals.
- **Zinc**: Supports immune function and wound healing. Sources include meat, shellfish, and legumes.
- **Magnesium**: Vital for many biochemical reactions. Sources include nuts, seeds, and leafy greens.

5. Probiotics:

- **Description**: Beneficial bacteria that can support gut health.
- **Use**: May help improve digestive health and manage symptoms. Consider supplements or probiotic-rich foods like yogurt and kefir.

Recommendations:

• **Consultation**: Always consult with a healthcare provider or dietitian before starting any supplementation to tailor choices to individual needs and avoid potential interactions.

• **Regular Monitoring**: Periodic assessment of nutritional status and symptoms can help guide adjustments in supplementation and diet.

By incorporating appropriate supplementation and focusing on nutrient-dense foods, individuals with EPI can better manage their condition and maintain overall health.

Traveling With EPI

Traveling with Exocrine Pancreatic Insufficiency (EPI) requires some planning to manage symptoms and maintain a balanced diet. Here are some tips to help make traveling easier:

• **Research Food Options**: Look for restaurants or markets that offer suitable meal choices, such as lean proteins, healthy fats, and easily digestible carbohydrates.

• **Pack Snacks**: Bring portable snacks like low-fat yogurt, rice cakes, or granola bars to avoid hunger and ensure you have safe options available.

• **Travel Supply**: Pack enough pancreatic enzyme supplements for the duration of your trip, plus a little extra in case of delays.

• **Storage**: Keep enzymes in their original packaging and consider a travel-friendly container for convenience. Carry a water bottle to stay hydrated, especially during travel. Dehydration can exacerbate symptoms.

• Try to stick to regular meal times and portion sizes to help manage digestion and minimize discomfort.

• Avoid foods that you know may trigger symptoms, such as high-fat or fried foods, and be cautious with new foods.

- Consider keeping a food diary during your trip to track what you eat and any symptoms that arise. This can help identify triggers in unfamiliar settings.

- Don't hesitate to communicate dietary needs at restaurants or with travel companions. Most establishments can accommodate special requests.

- If traveling to a location with limited food options, you might consider bringing multivitamins or other supplements as advised by your healthcare provider.

- Familiarize yourself with local healthcare facilities in case of a medical issue. Carry a note from your healthcare provider regarding your condition and necessary treatments.

By planning ahead and being mindful of your dietary needs, you can enjoy your travels while effectively managing EPI symptoms.

CHAPTER FIVE
EPI And Pregnancy

Managing Exocrine Pancreatic Insufficiency (EPI) during pregnancy requires careful attention to nutrition and health. Here are some important considerations:

1. Nutritional Needs:

• **Increased Nutritional Requirements**: Pregnancy increases nutritional demands, including higher needs for calories, protein, vitamins, and minerals.

• **Balanced Diet**: Focus on a well-balanced diet rich in lean proteins, healthy fats, whole grains, fruits, and vegetables.

2. Monitor Fat Intake:

- Since individuals with EPI may have trouble digesting fats, work with a healthcare provider to determine the right amount of healthy fats and consider pancreatic enzyme replacement therapy (PERT) to aid digestion.

3. Vitamins and Minerals:

- **Supplementation**: Consult with a healthcare provider about prenatal vitamins, especially for fat-soluble vitamins (A, D, E, K) and other essential nutrients, such as folic acid and iron.

- **Monitor for Deficiencies**: Regularly assess nutritional status to prevent deficiencies, which can affect both maternal and fetal health.

4. Hydration:

• Stay well-hydrated, as hydration is crucial during pregnancy. This can help with digestion and overall health.

5. Frequent, Small Meals:

• Eating smaller, more frequent meals can help manage symptoms and support better digestion.

6. Regular Monitoring:

• Regular check-ups with healthcare providers are essential to monitor both maternal and fetal health, as well as to adjust dietary and supplementation needs as pregnancy progresses.

7. Communicate with Healthcare Providers:

• Keep all healthcare providers informed about EPI and any symptoms that arise. They can provide tailored advice and support throughout pregnancy.

8. Be Aware of Symptoms:

• Watch for any changes in symptoms and discuss them with a healthcare provider, as pregnancy can sometimes affect digestive health.

With appropriate management and support, individuals with EPI can have a healthy pregnancy. Always work closely with healthcare providers to ensure optimal care.

Managing EPI In Children

Managing Exocrine Pancreatic Insufficiency (EPI) in children requires a comprehensive approach that focuses on nutrition, monitoring, and support. Here are some key strategies:

1. Pancreatic Enzyme Replacement Therapy (PERT):

- **Enzyme Supplements**: Administer prescribed pancreatic enzyme supplements with meals and snacks to help with the digestion of fats, proteins, and carbohydrates.

- **Dosage Adjustments**: Work with a healthcare provider to adjust enzyme dosages based on dietary intake and symptoms.

2. Balanced Diet:

• **Nutrient-Dense Foods**: Focus on a diet rich in whole foods, including lean proteins (chicken, fish, eggs), healthy fats (avocado, nuts, olive oil), and complex carbohydrates (whole grains, fruits, vegetables).

• **Limit Processed Foods**: Minimize intake of high-fat, fried, or heavily processed foods that can exacerbate symptoms.

3. Frequent, Small Meals:

Meal Frequency: Offer smaller, more frequent meals and snacks to ease digestion and prevent discomfort.

4. Monitor Growth and Development:

- **Regular Check-ups**: Monitor growth and development closely, as EPI can affect nutrient absorption and overall health. Regular assessments can help identify any nutritional deficiencies early.

5. Nutritional Supplements:

- **Vitamins and Minerals**: Consider supplementation for fat-soluble vitamins (A, D, E, K) and other nutrients as needed, based on healthcare provider recommendations.

- **Hydration**: Ensure adequate fluid intake to support overall health and digestion.

6. Education and Support:

- **Inform Caregivers**: Educate teachers, daycare providers, and other caregivers about the condition and any dietary restrictions or needs.

- **Create a Supportive Environment**: Encourage children to be involved in meal planning and preparation when appropriate, fostering a positive relationship with food.

7. Monitor for Symptoms:

- **Track Symptoms**: Keep a diary of food intake and symptoms to help identify any triggers or patterns that may indicate dietary adjustments are needed.

8. Collaboration with Healthcare Providers:

- **Multidisciplinary Approach**: Work closely with pediatricians, dietitians, and gastroenterologists to tailor dietary and treatment plans to the child's specific needs.

By implementing these strategies and maintaining open communication with healthcare providers, parents can effectively manage EPI in children and support their overall health and well-being.

Conclusion

Managing Exocrine Pancreatic Insufficiency (EPI) involves a multifaceted approach that emphasizes dietary modifications, appropriate supplementation, and regular monitoring. Individuals with EPI, including children and pregnant women, can lead healthy lives by:

- **Utilizing Pancreatic Enzyme Replacement Therapy (PERT)** to aid in digestion.
- **Focusing on a balanced diet** rich in lean proteins, healthy fats, and complex carbohydrates.
- **Monitoring growth and nutritional status** to prevent deficiencies, especially of fat-soluble vitamins.

- **Adapting meal patterns** to smaller, more frequent meals for better digestion.
- **Maintaining open communication** with healthcare providers to adjust treatment and dietary plans as needed.

By adopting these strategies and fostering a supportive environment, individuals with EPI can effectively manage their symptoms and improve their overall quality of life.

www.ingramcontent.com/pod-product-compliance
Lightning Source LLC
Chambersburg PA
CBHW070206230526
45471CB00002B/845
9798336479492